The Eildon Hills

Contents

EDITORIAL TEAM:

Malcolm Lindsay, David Parkinson, Lawrence Robertson and Louise Wallace

Published by
THE SCOTTISH WILDLIFETRUST CENTRAL BORDERS GROUP

First Edition 1976   Third Edition 1996
Second Edition 1987   Fourth Edition 2016

FOREWORD

I am delighted to recommend this little booklet, so lovingly prepared by the Central Borders Group of SWT. I hope that this introduction to the geology, plant life, animal life and historical and literary associations of the Borders will arouse your interest, if you are local, inspire you to learn more, and if a transient visitor, to entrance and entertain you with the insights and knowledge gathered in these few pages.

You will learn how the rocks and earth beneath your feet determine the plants and animals that can survive, how they relate to each other and to man, and have a feel for how landscape and Flora and Fauna influence rural culture, and the writings of such as the Ettrick Shepherd and Sir Walter Scott of great fame. It is as full of little delights as a Selkirk Bannock!

Be inspired to look around at all you see on a walk in the quiet beauty of the Borders countryside .Wherever you go you will be treading in someone's footsteps – St. Cuthbert on his way from Melrose to Holy Island across the Eildons, Sir Walter on a walk by the Tweed, A Border Reiver astride his horse on a cattle raid over the hills- history is in every blade of grass and every puff of wind!

Maybe it could play a part in winning the hearts of all the new visitors who arrive on the new railway line , deepening their understanding, and whetting their appetites for further visits

Well done to all of you who have over time been associated with this project and especially to those involved in the new edition.

It will be a great success, I am sure.

Robin Harper
Chairman, Scottish Wildlife Trust.

Welcome to the Eildons

We who live in the Borders have a deep affection for our "Eildons", ancient, familiar, presiding over their land, witnesses to histories beyond our fathoming, inspiring writers and artists, and giving a unique home to plants and wildlife.

There are four Eildon Hills, North, Mid, Wester, and the aptly named Little Hill. Even though the highest, Mid Hill, is only 422 metres, together they form the striking centre-piece of Eildon and Leaderfoot National Scenic Area and dominate the wide surrounding landscape.

Particularly dramatic views of the Eildons are obtained from the south and east – from the A68 road summit at Carter Bar on the Scotland-England border and from "Scott's View", a panoramic beauty spot loved by Sir Walter Scott high on Bemersyde Hill, above Dryburgh. You can see the Eildons, sometimes all three main hills, sometimes two, from many parts of the Borders. The Romans would have used them to navigate and we still do.

This means that views from the summits are spectacular. Looking northwards we see the Tweed Valley basin with its towns and villages, Galashiels, Tweedbank, Gattonside, Melrose and Newstead.

A little beyond to the immediate northeast the volcanic Black Hill rises beside Earlston. The backdrop of the northern vista is formed by the Moorfoot Hills to the west and the Lammermuir Hills to the east. To the south, we see the villages of Newton St Boswells, St Boswells and Bowden just below us with the prominent Waterloo Monument in the middle distance. The background vista is formed by the high Tweedsmuir Hills in the west, the distant hills of Liddesdale and Eskdale to the south and the summits of the Cheviots to the east. Scott said: *"I can stand on the Eildon Hill, and point out forty-three places famous in war and verse".* Well worth the climb.

The summit of Mid Hill is marked with a granite block carrying a large brass plate inscribed with the names of the many hills and other points of interest visible round the compass. This was erected in 1927 and bears the dedication –

"To the memory of Sir Walter Scott.
From this spot he was wont to view
and point the glories of
the Borderland".

By the Tweed, just to the west of the Eildons, Scott built Abbotsford, his home. Of it he said *"my heart clings to the place I have created".* His heart and his writings were also firmly rooted in his beloved Borders landscape, history and legends.

EILDON LEGENDS

People have always been in awe of the Eildons. There are several legends associated with the hills. It is said that there was originally a single large peak but this was cleft into three by a blow from the spade of the Devil who had been summoned by the wizard Michael Scott. The Devil wiped his spade making a little hillock -"The De'il's Spade Fu'"- which we now see as Little Hill, the knoll situated between Mid Hill and Wester Hill.

A more gentle story is that of Thomas the Rhymer, a 13th century writer and traveller, who met with the Queen of Elfland at the Eildon tree. She spirited him away to a fairyland inside the hills for seven years and the knowledge he gained there led to his brilliance as a poet and soothsayer. Another tale speaks of King Arthur and his knights and their steeds standing in a cavern within the hills, ready to ride out on Judgment Day.

The Eildon legends are used in a number of Scott's works including "The Eve of St John" (1799), "Minstrelsy of the Scottish Border" (1803) and "The Lay of the Last Minstrel" (1805).

The Ettrick Shepherd, James Hogg, also used the Michael Scott story in his novel "The Three Perils of Man" (1823).

Images of the Eildons

Artists and photographers have long been attracted to the Eildons intent on capturing their beauty and presence. Perhaps most prominent of these was the great landscape artist, John Mallord William Turner, who worked with Scott over several years and who stayed with him at Abbotsford in 1831. Several of Turner's watercolours depict the Eildons. They are part of the Vaughan Bequest which is displayed in the National Gallery of Scotland in Edinburgh every January. More recently canvasses by Earl Haig of Bemersyde have celebrated the Eildons and surroundings.

Some History

For as long as people have made their homes in this area the Eildon Hills will have been a familiar reassuring presence. The Eildons sit at the point where the Tweed and its tributaries break free of the constraining hills and take a slower course towards the sea, nurturing a large area that has been developed over centuries into good settled arable farmland. They also form a potent landmark seen from long distances on the approach to the area and command the kind of views that make them indispensable sentry points during warfare.

From around the first millennium BC farms and settlements were gradually wrought from the mixed forest and scrub that covered the area. The Roman road, Dere Street, made a bee line from the south to the Eildons, crossing the Tweed at the strategic military base known as Trimontium near present day Newstead. The Romans placed a signal station on Hill North. Archaeology suggests this would have been a wooden rectangular building inside a small circular ditched enclosure.

▲ *Eildon Hill Fort* © *Historic Scotland. Licensor www.scran.ac.uk*

The indigenous people, of whom we know little until we hear of the Selgovae and the Votadini in Roman times, have left evidence of considerable engineering in the area as ditches and earthworks throughout the Borders. Eildon Hill North has the largest hill fort in Scotland. The broad and fairly level summit has terraces on all sides, the most discernible being pierced with five entrances. Inside the ramparts are the remains of over 290 house platforms. Medieval cultivation and later woods may have obscured many more on a large area of level ground further down the southern slope, known as The Floors. Excavation revealed that the hill top site may have been in use from a thousand years before the Romans arrived. The reason remains a mystery.

As there is no water supply to support a permanent settlement, many of the buildings might have been for storage, or it may have

Wester Hill, Mid Hill and iron age rampart from Eildon Hill North

accommodated large gatherings for significant events. Ritual activity may be inferred by finds of Bronze Age axes at the base of the hills near springs that in time were later adopted as holy wells. The Siller Stane, an isolated large boulder situated near The Saddle between Hill North and Mid Hill, is so-called because of a belief that there was money under it. Unconfirmed suggestions have been made that it is a bronze age relic.

If earlier people imbibed a sense of magic from the Eildons this feeling continues through Medieval writings and even those of Sir Walter Scott.

A 12th century record gives the name Eldune, possibly derived from the Gaelic aill-meaning rock and dun-hill. An alternative theory suggests it to be a corruption of the Cymric moeldun, bald hill. Owners have come and gone but few have made significant changes to the landscape of the hills. The most visible man-made feature today is the woods around Eildon Hall, planted by Thomas Mein, owner in the early nineteenth century.

WHAT THE ROCKS TELL US

Several hundred million years ago the area around Melrose was covered with layer upon layer of sand and mud turning to rock. These "Silurian" rocks were then caught between moving continental plates and pushed together until mountains were thrown up. Ages of erosion followed until the mountains were pretty well flattened and their remains had sunk beneath the sea. Then they were covered with a deposit that was to become Old Red Sandstone which gives the characteristic red soils of the surrounding area.

Later there was much volcanic activity in the area, including the formation of the volcano that is now Cheviot, and other well-known Border landmarks visible from the Eildons such as Ruberslaw, and the Minto Hills. A considerable thickness of soft volcanic material would have been deposited over the Old Red Sandstone though little of this remains today.

Towards the end of this period of activity Melrose became the

▲ *400 million years in the making: Red sandstone visible at this field to the east of the Eildons.*

centre of a group of volcanoes with the nearest crater centred a mile to the north-west of the Eildons. Not all the molten rock reached the surface as lava and some, after breaking through the Old Red Sandstone, opened up and filled

How the Eildons were formed

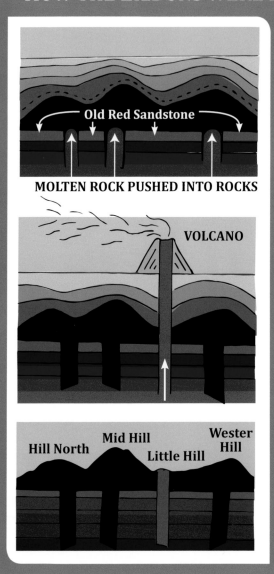

Old Red Sandstone

MOLTEN ROCK PUSHED INTO ROCKS

VOLCANO

Hill North · Mid Hill · Little Hill · Wester Hill

300 Million Years ago, The Eildons are formed beneath the surface of the Earth

250 Million years ago, Little Hill is a volcano

Millions of years of deposition and erosion reduce the surface layer

1 million years ago the ice sheets begin to grind down the hills to their present shape

a dome-shaped layer between the sandstone and the overlying deposits. The main mass of the Eildons was formed in this way, not all at once, but by successive intrusions of molten rock. This hard igneous rock has weathered much less than the surrounding softer sandstones. Later eruptions threw up ash and blocks of rock that can still be seen in layers on the hillsides.

Nearby, the Black Hill of Earlston was formed in a similar way. A late volcanic action formed Little Hill, partly filled by a solid plug of hard basalt. The basalt differs chemically from the other rocks and now carries a distinct vegetation.

We now jump forward in time across millions of years when the rocks over the Eildons were slowly eroding away, to the Ice Age that ended only about ten thousand years ago. The glaciers that covered the area flowed towards the north-east and as they moved they scraped layer after layer from the rocks over which they passed and only the hardest remained. The hard rocks of Mid Hill and Wester Hill are the principal reasons for the existence of the Eildon Hills today and even these are probably only a fraction of their original size. The slightly softer rock of Hill North has been smoothed on the north side by the glacier action as the ice flowed around the edge of Mid Hill. The softer Old Red Sandstone only survives where it is underneath the hard rocks.

Elsewhere the more ancient Silurian rocks have been exposed to give the rocky knowes we can see on the west and south slopes. The glacier left a tail of loose rocks on the north east side. The tail can be easily seen as we approach from the south. It can also be studied on the banks of The Tweed below Monksford House as the river cuts through huge blocks. In winter the bare fields to the east are red due to the high content of Old Red Sandstone.

Since the Ice Age frost action has splintered the steep rock faces to give the scree slopes we see today. Elsewhere the rock has weathered enough to create a thin soil that supports the vegetation and slows further erosion.

Places and Paths

The hills are covered by a very good network of paths including the long distance trail, St Cuthbert's Way, which runs from Melrose to Lindisfarne on the Northumberland coast. A good point of access to the Eildons is on St Cuthbert's Way in Dingleton Road 250m south of Melrose town centre (NT547338). Other access points are from the track leading from the old St Boswells road just east of Melrose cemetery (NT563336), from Eildon Mains just north of Eildon village (NT571329), from the track running from Melrose Golf Clubhouse (NT543332), from Bowdenmoor Reservoir (NT539316) and from the village of Bowden (NT554305). Excellent descriptions of paths and routes are provided in the booklet "Paths around Melrose" published by the Scottish Borders Council.

The hills are easily accessed from the terminus of the Borders Railway at Tweedbank. It is a walk of 2.5 Km from there to Melrose but a frequent bus service is available.

Path leading to Wester Hill

Walking on the hills is generally easy though there are some steep and occasionally muddy slopes on the summit approaches. Children love climbing the Eildons and obtain a great sense of achievement from the brief steep climbs leading to such well-defined summits. For very fit and adventurous hill users other activities include paragliding from Hill North, the Jedburgh Three Peaks Ultramarathon and the Eildon hill-running race which has been held annually in June for over 50 years.

▲ *Bowdenmoor Reservoir*

▲ *Between Mid-Hill and Wester Hill*

▲ *The Eildons in Winter*

▲ *A view from Eildon Hill North showing paths on Mid Hill and Wester Hill*

▲ *View from Bowden Road near Newtown*

▲ *Mid Hill and Siller Stane*

▲ *View from Cauldshiels Hill*

15

THE EILDON HILLS

- ☐ Open Hill
- ▦ Woodland
- ○ Access point to the hills
- ----- Paths
- ✠ St Cuthbert's Way
- ⩍ Eildon Hills path
- ⋇ Viewpoint
- ⚲ Golf course
- ℗ Parking
- ⁊⁊⁊ Scree
- ∩ Quarry
- ⊙ Spring/well
- ⌇ Damp areas

B 6359

Bowdenmoor
Res.

◀ *The Eildons from Scott's View*
in autumn

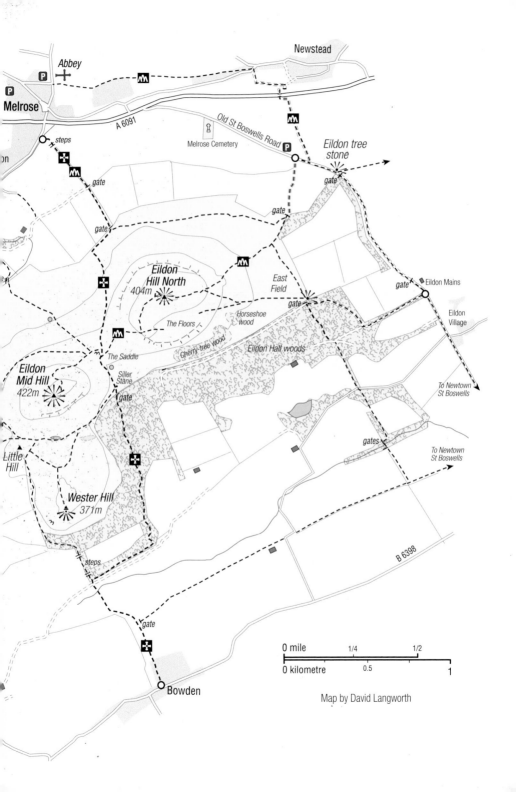

Newstead

Abbey

P

Melrose

P

A 6091

Old St Boswells Road

Melrose Cemetery

Eildon tree stone

steps

gate

gate

gate

gate

gate

Eildon Hill North
404m

East Field

gate

Eildon Mains

gate

Eildon Village

The Floors

Horseshoe wood

Cherry-tree wood

Eildon Hall woods

The Saddle

Eildon Mid Hill
422m

Siller Stane

gate

To Newtown St Boswells

Little Hill

Wester Hill
371m

gates

To Newtown St Boswells

steps

B 6398

gate

Bowden

0 mile   1/4   1/2

0 kilometre   0.5   1

Map by David Langworth

▲ *Harebells*
(Campanula rotundifolia)

FLOWERING PLANTS

The vegetation of the Eildons has developed over the millennia since the Ice Age in a way that reflects the underlying geology and the influence of man. The glaciers of the Ice Age sculpted the hills into the shape we see today. The west side is steep and there is much open rock scree and such soils as are present are very shallow 'skeletal' soils. They are dry soils as rainfall runs off very easily. The north and south sides are only a little less steep. The east side is different. To the east of Hill North there is a long slope that shallows out lower down. Here the glaciers deposited some of the material that had been scraped off the hills. The drainage is not so good and there are some wet places.

Most of the igneous rocks are acidic and the vegetation contains only a few species. Little Hill has rock that is slightly calcareous and a few specialist species find a home there. Both the Old Red Sandstone and the Silurian rocks of the lower slopes are more neutral but the soils of the lower slopes are either shallow or contain material scraped off the hill above, so they are of low fertility.

The springs feeding the wet places are concentrated at a 'spring line' where the igneous rocks of the upper part of the hills meet the Old Red Sandstone and again where the sandstone meets the Silurian rocks below. The springs on the north side, especially towards the west near the former rifle range above the golf course, are slightly calcareous and have a richer flora than that of the more acidic springs in East Field. There is a still more acidic spring at the east side of the col between Mid Hill and Wester Hill with a characteristic flora.

There is no trace on the Eildons of the woodland that covered much of Britain after the last Ice Age. This woodland would have covered the lower slopes but the upper slopes would have had much open heather moorland, just as today. The woodland has all been cleared by man leaving heather moorland on the upper slopes and hill grassland lower down. East Field has been cultivated in the past and the old 'rigs' can still be traced. It reverted to hill grassland after cultivation was abandoned except where whins (gorse) seized the opportunity to gain a hold. There are two modest blocks of woodland on the Eildons: the larger is 'Cherry-tree wood' near the Eildon Hall boundary. It was originally planted as a fox covert with a rather curious mixture of species including Gean (Wild Cherry) and Sloe (Blackthorn) but a wider variety of trees have now naturalised there. Almost adjacent to Cherry-tree wood is 'Horseshoe wood', a group of fine old Scots Pines planted in the shape of a horseshoe.

Studies done in the 1960's and early 1970's allow us to compare and observe recent development of the flora. The hills have now been managed as rough grazing for centuries. There has been no deliberate 'muirburn' for many years. Sadly, there have been several recent unplanned heather fires that have scarred the west side of the hills. The vegetation and associated wildlife is steadily recolonizing, but it is a slow process.

In 1974 there were still cattle and horses on the hills as well as sheep. Now there are only sheep. The

▼ *Mid Hill, August*

management policy is to keep grazing at a modest level to allow the heather to flourish for the benefit of wildlife.

At one time some of the whins may have been cut and crushed for use as winter feed for cattle

▲ *Path to Hill North bordered by gorse*

and horses and these animals will have helped to contain the whins in summer by grazing the young growth. Nevertheless the whins had colonised 80% of East Field by 1964 and they were then sprayed to control them. They were indeed controlled, but only for a time and by 1974 they had recolonized considerably. They have since spread further and now approach the coverage seen in 1964. Bracken has also spread in the grassland, but its spread has been limited by the shallowness of the soils.

Apart from the fires and the spread of whins, the most significant change in recent years has been the spread of trees and shrubs made possible by the low stocking density of the sheep. Scots Pine is the most prominent colonist of the heather moor, but there is also Larch, Sitka Spruce, Birch and Rowan. The Sitka has the potential to spread out of control. Meanwhile there are gaps among the whins which sheep do not penetrate, allowing colonisation by Hawthorn, Elder, brambles and wild roses.

The vegetation of the Eildon Hills has only modest species-diversity despite the variety of habitats. In a way the limited number of species enhances the attraction of the hills as the simplicity is appealing. A survey in 1974 found 224 species, fewer than could be found along the banks of the River Tweed. Since 1974 just a few species have been lost and there has been a similar number of new colonists.

Heather moorland

When viewed from a distance the Eildons appear to be all heather-covered. As one walks the hills one climbs through a band of grassland before reaching the heather. Not all the heather is Common Heather *Calluna vulgaris*: in places it is replaced by Bell Heather *Erica cinerea*, especially on the south side of the hills. The white-flowered form of heather is occasional, but any white-flowered Bell Heather must be very rare.

Near the summit of Mid Hill there is a good colony of Cowberry *Vaccinium vitis-idaea* which is a more northern species that keeps its leaves in the winter. It is also to be found in small quantity on Wester Hill. The more familiar Blaeberry (Bilberry) *Vaccinium myrtillus* is quite widespread but is especially prominent on the east slopes of Hill North where it is accompanied by Wood Anemone *Anemone nemorosa*, though that species is also found on the lower slopes. Great Wood-rush *Luzula sylvatica* is another woodland species that is frequent in heather on the east slope of Hill North but scarce elsewhere.

Hard-fern *Blechnum spicant* is the fern most characteristic of the heather moor: it has two sorts of frond, a tall skimpy one bearing the capsules which hold

1. Bell Heather and Common Heather *(Erica cinerea and Calluna vulgaris)*
2. Cowberry *(Vaccinium vitis-idaea)*
3. Blaeberry *(Vaccinium myrtillus)*
4. Wood Anemone *(Anemone nemorosa)*

21

1. Bitter-vetch (Lathyrus linifolius)
2. Tormentil (Potentilla erecta)

3. Parsley Fern
   (Cryptogramma crispa)
4. Wood Sage
   (Teucrium scorodonia)
5. Climbing Corydalis
   (Ceratocapnos
   claviculata)

the spores and a broader lower-growing one without such capsules. Heath Bedstraw *Galium saxatile*, Tormentil *Potentilla erecta* and Bitter-vetch *Lathyrus linifolius* are three grassland species that are also common in the heather moorland.

Rock screes

The nature of screes is to be unstable and for the loose rock to slip downhill over the years, so they are colonised by vegetation only with difficulty. Plants find a foothold behind one of the larger blocks of rock and there a community may develop precariously. Specialists of this habitat are Wood Sage *Teucrium scorodonia*, Foxglove *Digitalis purpurea* and Parsley Fern *Cryptogramma crispa*; the last is a speciality of the Eildons (though it is quite frequent further west towards Peebles). It is another fern with two sorts of frond only one of which is fertile. Climbing Corydalis *Ceratocapnos claviculata* is an attractive scrambling annual plant that occurs in modest quantity on scree below Wester Hill and at the woodland edge nearby. Rosebay Willowherb *Chamerion angustifolium* occurs on screes but is especially frequent in the small quarries and among the whins. Lemon-scented Fern *Oreopteris limbosperma* prefers slightly damp places but may occur near the foot of screes and below rocks.

Little Hill

Little Hill has a more grassy vegetation than the other hills due to the more calcareous nature of the rock. Its specialities include two favourites of the Scottish Borders, Common Rockrose *Helianthemum nummularium* on the south-facing side and Mountain Pansy *Viola lutea* on the north-facing side. Kidney Vetch *Anthyllis vulneraria* is a lowland species present here in small quantity with some small annual species that are quite scarce in the area.

Grassland

The most floriferous parts of the grassland on the lower slopes are the dry banks. Typical species are Common Bird's-foot-trefoil *Lotus corniculatus*, Wild Thyme *Thymus polytrichus*, Harebell *Campanula rotundifolia* and Heath Speedwell *Veronica officinalis*. Lady's Bedstraw *Galium verum* and Heath Milkwort

1. *Common Rockrose (Helianthemum nummularium)*
2. *Common Bird's-foot-trefoil (Lotus corniculatus)*
3. *Heath Milkwort (Polygala serpyllifolia)*
4. *Heath Speedwell (Veronica officinalis)*

1. Wood Sorrel
   (Oxalis acetosella)
2. Common Spotted
   Orchid *(Dactylorhiza
   fuchsii)*
3. Cuckooflower
   (Cardamine pratensis)
4. Ragged Robin
   (Lychnis flos-cuculi)

Polygala serpyllifolia are a little less frequent. Slender St John's-wort *Hypericum pulchrum* and Common Dog-violet *Viola riviniana* prefer slightly shady places. Bracken *Pteridium aquilinum* is dominant on some of the sheltered slopes where Wood Sorrel *Oxalis acetosella* is plentiful beneath its fronds.

Wet places

The most interesting area for wetland species is a series of small springs on sloping ground in the area above the burn that adjoins the golf course and around the old rifle range targets nearby. These springs are calcareous and are rich in sedges. Other species present include Quaking-grass *Briza media*, two similar orchid species, Common Spotted-orchid *Dactylorhiza fuchsii* and Northern Marsh-orchid *D. purpurella*, Marsh Lousewort *Pedicularis palustris*, Cuckooflower *Cardamine pratensis,* Ragged Robin *Lychnis flos-cuculi,* Common Butterwort *Pinguicula vulgaris* and a very little Grass-of-Parnassus *Parnassia palustris*. Devil's-bit Scabious *Succisa pratensis* is more widespread.

A third orchid, Heath Spotted-orchid *Dactylorhiza maculata*, is to be found with Cottongrass *Eriophorum angustifolium* in the more acidic spring at the east side of the col between Mid Hill and Wester Hill. There are various springs and runnels in East Field but they have few notable species.

Bowdenmoor Reservoir

The track to the Eildons from the west starts at Bowdenmoor Reservoir. This is surrounded by a plantation of mature trees. The fringe of the water area has much emergent Water Horsetail *Equisetum fluviatile,* Water Mint *Mentha aquatica* and rafts of Amphibious Bistort *Persicaria amphibia.* White Water-lily *Nymphaea alba* and Yellow Iris *Iris pseudacorus* flower prominently from early summer. The waterlilies with yellow flowers are not the native Yellow Water-lily *Nuphar lutea* but an introduction, Fringed Water-lily *Nymphoides peltata*, which is related to Bogbean *Menyanthes trifoliata.*

Losses and colonisation

The most notable losses since 1974 have been to the clubmosses.

1. *Water Mint (Mentha aquatica)*

2. *Amphibious Bistort (Persicaria amphibia)*

3. *Yellow Iris (Iris pseudacorus)*

4. *White Water-lily (Nymphaea alba)*

Stag's-horn Clubmoss *Lycopodium clavatum (pictured right)* was formerly found in several places on the Eildons, but it has been reduced to an occasional casual specimen, mainly near the summit of Mid Hill. Eutrophication related to the atmospheric deposition of nitrogen compounds may be to blame.

Of the new arrivals perhaps the most interesting is New Zealand Willowherb *Epilobium brunnescens*, found at Wester Hill in 2012 on a damp recently-burnt slope. This is a species which has colonised all the upland areas of Britain in the last 50 years. It is a low-growing plant that merges in well with the native vegetation.

FUNGI, LICHENS AND BRYOPHYTES

In south-east Scotland, basaltic hills are often rich in bryophytes different from those of the surrounding lowlands. This is true for the Eildons, though the three main hills are composed of quiteacidic rocks and dominated by heathy vegetation, which tends to support mostly common species. In October 2015 a bryophyte survey gave the respectable total of 83 different mosses and 24 liverworts.

The most interesting area is Little Hill, a volcanic vent with strongly calcareous rocks and rich grassland. On its rocky outcrops are found two local mosses, Spreading-leaved Grimmia *Grimmia curvata* and Bird's-foot Wing-moss *Pterogonium gracile* and the leafy liverwort Spotty

Scalewort *Frullania fragilifolia.* and nearby the large Sea Ivory lichen *Ramalina siliquosa*, normally growing on rocks by the sea and rare inland.

Another special area is the small valley below the south slope of

Mid Hill, with calcareous flushes supporting five *Sphagnum* species and the beautiful Golden-head Moss *Breutelia chrysocoma,* not recorded on the Eildons since 1868. Another moss with the same historical gap is Donn's Grimmia *Grimmia donniana* found on small stones near the summit of Mid Hill.

The Eildons are rich in fungi, and two attractive species present on the grassy slopes of the Little Hill are Yellow Club *Clavulinopsis helvola* and the Dusky Puffball *Lycoperdon nigrescens.*

1. Golden-Head Moss *(Breutelia chrysocoma)*
2. Yellow Club Fungus *(Clavulinopsis helvola)*

BIRDS

The bird species seen in the area are typically those of the open Southern Upland hills but the species list is considerably extended by the presence of water at Bowdenmoor reservoir and extensive woodland, both deciduous and coniferous, on the southern aspect of the hills.

In spring Wheatears, summer migrants from Africa, are often seen around Little Hill but breeding on the hills has not been proved. Cuckoos are always heard on the lower slopes from the first week of May. The return

of two or three pairs is welcomed by us but not by the Meadow Pipits *(left)* which can often be seen mobbing a Cuckoo. Their concern is well-founded as the Cuckoo most frequently lays in Meadow Pipit nests.

1. Willow
   Warbler
2. Yellowhammer
3. Red Grouse
4. Buzzard

Other early summer arrivals are Whitethroats, which breed in the gorse, and Blackcaps and Spotted Flycatchers, which favour the deciduous woodland on the south side of Eildon Hill North. Common hedgerow birds such as Blackbirds, Wrens, Dunnocks, Linnets and Yellowhammers are found in the field margins and in the gorse. In the deciduous woodland common species include Willow Warbler, Chiffchaff, Robin, Great Tit, Blue Tit, Coal Tit, Treecreeper and Great Spotted Woodpecker. Nuthatches, a recent coloniser of the Borders, will also be found here. Around the Wester Hill the coniferous woods have breeding Siskins and Crossbills which may be seen feeding on pine cones.

On the hills the commonest small bird is the Meadow Pipit with Buzzards and Ravens often seen flying over. Until recently a small colony of Red Grouse was present throughout the year, feeding on heather shoots, and hopefully they will recolonize the hills from nearby heather moors. The Stonechat was an emblematic bird of the Eildons until it was wiped out by the severe winters of 2009/10 and 2010/11 but a series of mild winters should herald its return.

Kestrels *(left)* and Sparrowhawks breed in the area every year. Kestrels are most frequently seen hovering over the bare hillside scanning the ground for mice and voles, while Sparrowhawks are more likely to be seen in the woods or over open fields. During the summer, visitors may be surprised to find a nest or brood of Mallard. This common duck frequently nests well away from water, and after hatching, the brood will be led to the nearest stream to reach the relative safety of the Tweed.

Bowdenmoor reservoir has breeding Coots, Moorhens, Mallard and sometimes Mute Swans. In early spring Grey Herons are attracted by spawning frogs and toads. The reservoir is more interesting to visit in the autumn and winter when wintering wildfowl use Bowdenmoor as one of a chain of five small lochs in the area. The most regular species are Mallard, Tufted Duck, Goosander and Goldeneye with occasional visits from Whooper Swans and Cormorants. Geese seldom visit this site but both common species of grey geese, Greylag and Pink-footed, are regularly seen flying in the area.

In autumn and winter the hedgerows of hawthorn, elder and blackthorn and the rowan trees on the hill provide valuable food for wintering Blackbirds, Fieldfares and Redwings. If there are turnip or stubble fields close by it is worth looking out for large flocks of mixed finches feeding on the weed seeds. Gatherings of huge numbers of Rooks and Jackdaws are common in the late autumn and early winter. Even in the deepest snow, bird life still exists on the Eildons. Wrens and Dunnocks can be seen in the gorse beneath the snow, flocks of Yellowhammers and Linnets search for seeds and Bullfinches may be seen higher up feeding on heather seeds.

The diverse range of bird species in the area of the Eildons is indicative of the wide variety of habitat on or near the hills. In a very small area there is a representative selection of the South of Scotland's bird life.

Amphibians and Mammals

Few records are available from the Eildon Hills and even some of the commonest animals are shown only as 'probable' in the absence of reliable information.

Amphibians and Reptiles

In early spring the common frog *(right)* and common toad spawn in good numbers in Bowdenmoor Reservoir and are occasionally seen in damp places on the hills, particularly near the golf course burn. The smooth newt and great crested newt have also been recorded near the golf course burn. There are no records of common lizard on the Eildon Hills despite suitable habitat and similarly there have been no confirmed sightings of adders in the last forty years.

Mammals

Rabbits, brown hares *(right)*, foxes and roe deer may occasionally be sighted on the open hills and also badgers at dusk, but they are usually found on the lower slopes where there is cover provided by gorse and bracken. This habitat is home to moles, shrews, field mice, short-tailed voles and brown rats. These small rodents, together with rabbits, provide food for stoats and weasels which are sometimes seen on the lower slopes.

A recent increase in the otter population in the Scottish Borders means that there is a chance of encountering this fine animal particularly along the golf course burn where mink are occasionally seen. Grey squirrels are frequently sighted in the Eildon Hall woods and with luck you may spot red squirrels there. Along the edge of these woods is also the best place to observe pipistrelle bats at dusk and other species of bat are very likely to be present.

BUTTERFLIES AND MOTHS

The grassland habitats on the lower slopes of the hills hold the largest populations of butterflies and moths. These areas have vegetation of only modest species diversity and most of the butterflies and moths found there are common species. Rare species with specialist foodplant requirements are not well represented.

▲ *Scotch Argus*

Nevertheless the observant visitor to the lower slope grasslands is likely to encounter good numbers of butterflies and day-flying moths. This is especially so in the wet grasslands on and near the old rifle range adjacent to Melrose Golf Course. Here Orange-tips *Anthocaris cardamines* and Green-veined Whites *Pieris napi* flutter delightfully in spring and early summer while Ringlet *Aphantopus hyperantus*, Small Heath *Coenympha pamphilus*, Meadow Brown *Maniola jurtina*, Dark-Green Fritillary *Argynnis aglaja* and Common Blue *Polyommatus icarus* *(right)* butterflies are on the wing from June until August. In recent years the Small Skipper *Thymelicus sylvestris* has been spreading northwards arriving in the Borders in 2006. It can now be found skipping energetically from flower to flower in July and August. From early August the Scotch Argus *Erebia aethiops*, an exquisitely beautiful chocolate-brown butterfly whose UK distribution is almost entirely Scottish, flies over its larval foodplant, Purple Moor-grass *Molinia caerula* .

Day-flying moths also frequent these same grassy places from June to August. Species include Chimney Sweeper *Odezia atrata*, Latticed Heath *Chiasmia clathrata*, Straw Dot *Rivula sericealis*, Silver-ground Carpet *Xanthorhoe*

montanata and Shaded Broad-bar *Scotopteryx chenopodiata* as well as several species of grassland micro-moths. A small colony of the Narrow-bordered Five-spot Burnet moth *Zygaena lonicerae* *(right)* is present near the old rifle range. The striking

red and black adults fly in late June and July. Higher up on the heathery slopes the walker may come across several species of large day-flying moorland moths. In April and May males of the beautiful Emperor moth *Saturnia pavonia* dash rapidly over the slopes searching for the sedentary females. In May and June the Fox Moth *Macrothylacia rubi* is on the wing and a month later Northern Eggar *Lasiocampa quercus callunae* males zig-zag wildly in afternoon sunshine. The hairy caterpillars of these moths and of the nocturnal moth Dark Tussock *Dicallomera fascelina* may be found on heather in late summer, autumn and spring. Borderers call these caterpillars "hairy oobits". They are an important food source for the hillside cuckoos. Common Heath *Ematurga atomaria* (adults flying May and June), Northern Spinach *Eulithis populata* (July and August) and Vapourer *Orgyia antiqua* (September and October) are other day-flying moths of the heathery tops.

The highly colourful, strong flying Vanessid butterflies, Small Tortoiseshell *Aglais urticae*, Peacock *Inachis io*, Red Admiral *Vanessa atalanta* and Comma *Polygonia c-album* can be seen anywhere on the hills. These species are mostly seen in spring after winter hibernation and again in August and September when their summer broods emerge. In exceptional years they are joined by large numbers of Painted Lady *Cynthia cardui* butterflies migrating from the south.

(left) Fox Moth caterpillar